T0224953

Cambridge Elements ⎓

Elements in Geochemical Tracers in Earth System Science
edited by
Timothy Lyons
University of California
Alexandra Turchyn
University of Cambridge
Chris Reinhard
Georgia Institute of Technology

SELENIUM ISOTOPE PALEOBIOGEOCHEMISTRY

Eva E. Stüeken
University of St Andrews
Michael A. Kipp
University of Washington

CAMBRIDGE
UNIVERSITY PRESS

CAMBRIDGE
UNIVERSITY PRESS

University Printing House, Cambridge CB2 8BS, United Kingdom

One Liberty Plaza, 20th Floor, New York, NY 10006, USA

477 Williamstown Road, Port Melbourne, VIC 3207, Australia

314–321, 3rd Floor, Plot 3, Splendor Forum, Jasola District Centre,
New Delhi – 110025, India

79 Anson Road, #06–04/06, Singapore 079906

Cambridge University Press is part of the University of Cambridge.

It furthers the University's mission by disseminating knowledge in the pursuit of
education, learning, and research at the highest international levels of excellence.

www.cambridge.org
Information on this title: www.cambridge.org/9781108749169
DOI: 10.1017/9781108782203

First published 2020

A catalogue record for this publication is available from the British Library.

ISBN 978-1-108-74916-9 Paperback
ISSN 2515-7027 (online)
ISSN 2515-6454 (print)

Selenium Isotope Paleobiogeochemistry

Elements in Geochemical Tracers in Earth System Science

DOI: 10.1017/9781108782203
First published online: September 2020

Eva E. Stüeken
University of St Andrews

Michael A. Kipp
University of Washington

Author for correspondence: Eva E. Stüeken, ees4@st-andrews.ac.uk

Abstract: The attraction of selenium isotopes as a paleoenvironmental tracer lies in the high redox potential of selenium oxyanions (Se^{IV} and Se^{VI}), the two dominant species in the modern ocean. The largest isotopic fractionations occur during oxyanion reduction, which makes selenium isotopes a sensitive proxy for the redox evolution of our planet. As a case study we review existing data from the Neoarchean and Paleoproterozoic, which show that significant isotopic fractionations are absent until 2.5 Ga, and prolonged isotopic deviations only appear around 2.3 Ga. Selenium isotopes have thus begun to reveal complex spatiotemporal redox patterns not reflected in other proxies.

Keywords: selenium isotopes, redox proxy, Earth evolution

ISBNs: 9781108749169 (PB), 9781108782203 (OC)
ISSNs: 2515-7027 (online), 2515-6454 (print)

Contents

1 Introduction   1

2 Materials and Analytical Methods   5

3 Proxy Mechanics   8

4 Case Study   12

5 Future Prospects   14

1 Introduction

Research on the geochemical behaviour of selenium (Se) goes back to at least the beginning of the twentieth century. Two key properties were recognized early on that have since driven the development and application of Se isotopes as a biogeochemical proxy: (1) Se can substitute for sulphur (S) in sulphide minerals because of its similarity in charge and radius, and (2) oxidation of selenide minerals and particulate elemental Se occurs at a higher redox potential than does sulphide oxidation (Figure 1), which leads to a relative suppression of Se mobility in the environment under moderately reducing conditions (Goldschmidt & Hefter, 1933). In addition, Se can occur in its elemental solid form over a wide pH range (Figure 1), which further inhibits its oxidation to soluble oxyanions. These attributes make Se a useful redox proxy that is archived in the geological record and has the potential to provide novel insights into the evolution of Earth's oceans and atmosphere (Stüeken, 2017).

Recent studies have provided a more nuanced picture of the biogeochemical Se cycle. Importantly, Se is an essential nutrient for some organisms (e.g., Stolz et al., 2006; Gladyshev, 2012) and therefore shows a nutrient-type profile in the modern open ocean (Cutter & Cutter, 2001). Such a profile means that concentrations are lowest in the photic zone where algal productivity peaks, and concentrations increase with depth during remineralisation of sinking organic matter. In anoxic basins, such as the Black Sea, this remineralisation pathway is suppressed, leading to depleted Se levels in the lower water column (Cutter, 1982). The reduction of the Se oxyanion selenate (SeO_4^{2-}), which is the most soluble form of Se, has been shown to occur at a similar redox potential as biological nitrate reduction (denitrification) (Oremland, 1990), that is, in suboxic regimes that are commonly encountered in oxygen minimum zones and in sedimentary pore waters. Fully oxic conditions (> 1 μM O_2) are therefore required to maintain a significant dissolved Se reservoir in the environment (Cutter, 1982; Rue et al., 1997). Collectively, these observations underpin the utility of Se as a redox proxy in deep time.

In accord with its aqueous behaviour, the major hosts of Se in the sedimentary record are unremineralized organic matter (including coal in terrestrial settings) and Se bound in diagenetic or syngenetic pyrite (Fan et al., 2011; Large et al., 2014; Stüeken et al., 2015c). In addition, Se oxyanions, in particular selenite (Se^{IV}), adsorb strongly to FeMn-oxides that form on the oxygenated seafloor (Mitchell et al., 2013). Broadly speaking, these three major sinks of Se from the ocean (organic matter, pyrite, FeMn oxides) are thus similar to those of molybdenum (Mo), where the mass balance of sinks has proved to be sensitive to the extent of ocean oxygenation (Kendall et al., 2017). A similar response can be

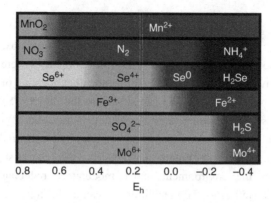

Figure 1 Stability of various chemical species as a function of redox potential (E_h)

Values were generated for $[X_i] = 0.1$ nM (where X is element of interest and a uniformly low concentration is utilized to avoid excessive precipitation of solid phases), standard temperature and pressure and a pH range of 7–8. Darker shading denotes progressively more reduced species. Oxidized selenium species (Se^{6+} and Se^{4+}) are only stable at relatively high redox potential; Se reduction begins at similar redox potentials to nitrate reduction. In contrast, Mo^{6+} and SO_4^{2-} are stable at lower redox potentials.

expected for Se (Section 2), although it is important to consider the significantly shorter residence of Se in seawater (10–20 kyr) compared to Mo (440 kyr), which probably makes Se a more regional rather than global redox proxy. Se is also cycled through the atmosphere in the form of elemental Se° particulates and methylated selenides (Wen & Carignan, 2007), analogous to biogenic methylated sulphides, which has stimulated ideas of using Se as a novel tracer for atmospheric evolution (Pogge von Strandmann et al., 2014; Stüeken et al., 2015a). However, the volatility of SeO_2 is much lower than that of SO_2, which may render it difficult to identify atmospheric signals in sedimentary archives.

While important insights have been gained from studies of Se abundances in pyrite through time (Large et al., 2014), a deeper understanding of transitions in the biogeochemical Se cycle can be gained from Se isotopes. Isotopic ratios can be used to fingerprint Se sources and reaction pathways in past environments. Se has six stable isotopes (Table 1), which experience mass-dependent fractionations of up to several permil during low-temperature reactions (Table 2). Se isotopic data are commonly reported in delta notation relative to NIST SRM3149, using either the $^{82}Se/^{78}Se$ or $^{82}Se/^{76}Se$ ratio. In the following, we will report isotopic data in terms of $\delta^{82/78}Se$ (Eq. 1).

Table 1 Stable selenium isotopes. Modified from Stüeken (2013) and Kurzawa et al. (2017) Note that ^{82}Se is considered stable for practical purposes because its half-life is $1.08 \cdot 10^{20}$ years.

Isotope	^{74}Se	^{76}Se	^{77}Se	^{78}Se	^{80}Se	^{82}Se
Exact mass [amu]	73.922	75.919	76.92	77.917	79.917	81.917
Natural abundance [%]	0.87	9.36	7.63	23.78	49.61	8.73
ICP-MS interferences	^{74}Ge, ^{36}Ar^{38}Ar, ^{58}Ni^{16}O ^{56}Fe^{16}O	^{76}Ge, ^{36}Ar^{40}Ar, ^{38}Ar^{38}Ar, ^{75}AsH, ^{60}Ni^{16}O	^{76}SeH, ^{40}Ar^{37}Cl, ^{36}Ar^{40}ArH, ^{38}Ar^{38}ArH	^{77}SeH, ^{38}Ar^{40}Ar, ^{78}Kr, ^{62}Ni^{16}O	^{40}Ar^{40}Ar, ^{80}Kr, ^{79}BrH, ^{64}Zn^{16}O, ^{64}Ni^{16}O	^{82}Kr, ^{81}BrH, ^{66}Zn^{16}O

Table 2 Low-temperature isotopic fractionations. The fractionation $\varepsilon_{R\text{-}P}$ between the reactant (R) and the product (P) of a biogeochemical reaction is defined as $\varepsilon_{R-P} = 1000 \cdot (\alpha_{R-P} - 1)$, where $\alpha_{R-P} = \left(^{82}Se/^{78}Se\right)_R \big/ \left(^{82}Se/^{78}Se\right)_P$. Modified from Stüeken (2017) and references therein

Pathway	$\varepsilon(^{82}Se/^{78}Se)$
Reduction	
abiotic Se(VI) → Se(IV)	5.6‰ to 11.8‰
abiotic Se(IV) → Se(0)	4.6‰ to 11.2‰
biotic Se(VI) → Se(IV)	0.2‰ to 5.1‰
biotic Se(IV) → Se(0)	1.1‰ to 8.6‰
(a-)biotic Se(0) → Se(-II)	< 0.5‰
Oxidation	
(a-)biotic Se(-II) → Se(0)	< 0.5‰
(a-)biotic Se(0) → Se(IV)	< 0.5‰
(a-)biotic Se(IV) → Se(VI)	< 0.5‰
Adsorption	
Se(IV) on FeMn-Oxide	<0.1‰
Se(VI) on FeMn-Oxide	< 0.7‰, average ~0.1‰
Volatilization	
Se(IV)/(VI) → CH$_3$-Se(-II)	2‰ to 4‰
Assimilation	
Se(IV)/(VI) → org. Se(-II)	<0.6‰

$$\delta^{82/78}Se[‰] = \left(\left(^{82}Se/^{78}Se\right)_{sample} \big/ \left(^{82}Se/^{78}Se\right)_{SRM3149} - 1 \right) \cdot 1000 \quad (1)$$

Isotopic fractionations between a product and a reactant will be reported in terms of ε values defined in Eq. 2.

$$\varepsilon = \left(\left(^{82}Se/^{78}Se\right)_{reactant} \big/ \left(^{82}Se/^{78}Se\right)_{product} - 1 \right) \cdot 1000 \quad (2)$$

The first analyses of Se isotopes were conducted by gas-source mass spectrometry on SeF$_6$ gas (Krouse & Thode, 1962), which is an established method for S isotope measurements. However, the relatively high sample requirement of >10 µg, in addition to the difficulty of quantitative conversion of all Se to SeF$_6$, hindered applications of this method to natural samples. The first geochemical measurements were done by thermal ionization mass spectrometry (TIMS), which lowered the sample requirement more than tenfold (Wachsmann & Heumann, 1992; Johnson et al., 1999). Since then, the development of multi-collector inductively coupled plasma mass spectrometry (MC-ICP-MS), combined with a hydride generator for sample introduction (see Section 2), has further boosted analytical sensitivity, allowing isotopic measurements on a few ng of Se (Rouxel et al., 2002; Kurzawa et al., 2017; Pons et al., 2020). This technological advance, paired with refinements of sample purification techniques (Clark & Johnson, 2010; Stüeken et al., 2013; Pogge von Strandmann et al., 2014; Kurzawa et al., 2017), has now opened up the possibility of investigating the geobiological Se cycle in full. In the following, we will provide a review of analytical conventions in Se isotope geochemistry and present a case study that illustrates the successful application of Se to a biogeochemical problem in Deep Time. The aim of this review is to demonstrate the utility of this emerging proxy in the Earth sciences.

2 Materials and Analytical Methods

2.1 Bulk Rock Analyses

Geochemical studies of Se isotopes most commonly focus on bulk rock analyses, which often involve complete dissolution of the rock matrix in a mixture of hydrofluoric acid and nitric acid (Rouxel et al., 2002; Stüeken et al., 2013). Perchloric acid may be added to oxidize recalcitrant organic matter and to ensure complete conversion of reduced Se to Se^{VI}. Other studies avoid the use of hydrofluoric acid and instead apply a leaching technique with nitric acid or inverse aqua regia (Mitchell et al., 2012; Kurzawa et al., 2017). In all cases, the subsequent solution is dried down and the residue is boiled in hydrochloric acid to convert Se^{VI} to Se^{IV}. The Se is then separated from other elements by chromatography. Several studies have used thiol cotton fibre (TCF), which can be prepared in-house and preferentially retains Se (Rouxel et al., 2002; Mitchell et al., 2012; Stüeken et al., 2013). However, TCF cannot fully separate Se from germanium (Ge) and arsenic (As) (Stüeken et al., 2013). To overcome this problem, Kurzawa et al. (2017) developed a method based on commercially available anion and cation exchange resins. Additional purification of Se can also be achieved by treatment with aqua regia, which removes Ge from solution (Stüeken et al., 2013).

After chromatographic purification, as established for analyses of rock matrices (see earlier), the Se-bearing hydrochloric acid is usually introduced into a hydride generator where Se^{IV} is converted to H_2Se gas by reaction with $NaBH_4$ (Rouxel et al., 2002). This step also provides additional separation of Se from Ge and organic compounds (Clark & Johnson, 2010; Banning et al., 2018). The sensitivity can be further enhanced by a factor of 2–3 if the H_2Se gas is mixed with CH_4 gas prior to introduction into the plasma torch (Kurzawa et al., 2017). Alternatively, Se can be introduced into the plasma with a desolvation nebulizer if the solution is doped with magnesium (Mg), which increases the sensitivity 100- to 200-fold compared to undoped desolvation nebulization (Pogge von Strandmann et al., 2014), although not necessarily relative to conventional hydride generation. In the mass spectrometer, Se isotopes suffer from numerous isobaric interferences involving a variety of elements (see Table 1). Earlier studies were able to circumvent this problem with a collision cell (Rouxel et al., 2002), which destroys ArAr dimers and other molecules in the plasma. This technique may be revisited in the future by other laboratories, in particular with the recent development of new collision cell multi-collector mass spectrometers by Nu Plasma and Thermo Fisher Scientific, but the technology is currently not widely available. Careful purification of Se during sample preparation and independent monitoring of Ge and As contents in the sample are therefore important. Masses 78 and 82 are least affected by interferences and therefore preferred in isotopic studies.

Isotopic fractionation within the mass spectrometer (instrumental mass bias) and temporal drift thereof can be corrected with either standard-sample bracketing or a double-spike. In the case of standard-sample bracketing, measured isotopic ratios of a sample are calibrated relative to the average of the two bracketing standards whose isotopic composition is known. This technique helps track isobaric interferences accurately by monitoring the proportionality between multiple isotopic ratios. It also allows detection of mass-independent fractionation (MIF), if present. In the case of a double-spike, two Se isotopes are masked by a spike with a known ratio, which has the advantage that isotopic fractionations induced during sample preparation can be corrected, leading to improved analytical precision. However, double-spiking only works when interferences on the spike isotopes are adequately corrected. Furthermore, this technique lacks the ability to detect mass-independent fractionation effects. Both techniques are therefore in use, depending on the application (e.g., Kipp et al., 2017; Labidi et al., 2018).

2.2 Phase-Specific Analyses

In addition to bulk rock analyses, a few studies have begun to investigate Se bound to specific mineral phases and organic matter and the results have revealed significant isotopic heterogeneity (Clark & Johnson, 2010; Schilling et al., 2014; Stüeken et al., 2015c). As noted earlier (Section 1), the major sinks of Se from the ocean are organic matter, incorporation into sulphide minerals and adsorption to FeMn-oxides. In addition, some elemental Se^0 may form, because it is thermodynamically stable at intermediate redox potentials (see Figure 1). Existing data show that all of these phases may coexist in a sedimentary rock, displaying distinct isotopic compositions. As the largest isotopic fractionations in the Se cycle are associated with dissimilatory Se oxyanion reduction (Table 2), Se^0 and sulphide-bound selenide tend to be isotopically depleted relative to organic Se and adsorbed oxyanions in the same sample (Stüeken et al., 2015c). This observation, although it is based on a very limited dataset so far, would imply that differential mixing of isotopically distinct Se phases can potentially produce significant variability in bulk rock measurements, which complicates the interpretation of bulk rock data. Phase extractions may therefore be important to fully reconstruct Se cycling in past environments. However, phase extraction protocols are so far not standardized and no suitable reference materials exist for inter-laboratory comparison. Furthermore, phase extraction procedures bear the risk of incomplete extraction from one phase, which may then affect subsequent phases and yield erroneous results. These problems need to be addressed with further method development.

2.3 Fidelity of the Record

As for other geochemical proxies, the three major processes that can perturb primary environmental signatures archived in the rock record are diagenesis, metamorphism and modern weathering. Diagenesis in sedimentary pore waters can mask signals produced by reactions in the overlying water column and it may homogenize isotopically distinct phases. However, it is important to note that significant isotopic fractionations require the availability of Se oxyanions for dissimilatory reduction. If Se isotopes are used as a proxy for ocean oxygenation, then diagenetic overprint would be part of the environmental signal. In other words, a significant diagenetic isotope perturbation would imply that the ancient seafloor was sufficiently oxygenated to maintain a reservoir of Se oxyanions in sedimentary pore fluids. Regarding isotopic homogenization during diagenesis, existing data suggest that this is not a major concern, because significant isotopic differences are preserved between different host phases in ancient rocks (Stüeken et al., 2015c). However, more samples need to be investigated to verify this conclusion, and modern sediment profiles should be studied to test how Se isotopes behave during burial. Metamorphism is also unlikely to cause shifts in Se isotopes, because laser-ablation analyses of Se bound in pyrite show negligible mobilization of Se with increasing metamorphic grade (Large et al., 2014). Se is probably too heavy to become volatile under these conditions.

Oxidative weathering of samples can potentially cause problems, as studies of soils have revealed large isotopic fractionations caused by re-reduction of oxidized Se (Zhu et al., 2014; Schilling et al., 2015). Furthermore, preferential oxidation of pyrite over recalcitrant organic matter may shift the composition of bulk rocks to heavier values, for example. It is therefore important to avoid analyses of weathered samples; for studies of ancient sedimentary rocks, this can be achieved by targeting well-preserved drillcore samples whenever possible.

3 Proxy Mechanics

3.1 Meteorites and Igneous Processes

Given the relatively low abundance of Se in igneous rocks (Stüeken, 2017), systematic studies of Se isotope fractionation in high-temperature processes have only recently begun. It is known that iron meteorites (~23 ppm) are enriched in Se compared to chondrites (~10 ppm) (Rouxel et al., 2002) and both are enriched relative to Earth's crust and upper mantle (both ~30 ppb) (Stüeken, 2017), which indicates that Earth's core likely sequestered large amounts of Se during planetary differentiation, leaving the silicate mantle and crust relatively depleted (König et al., 2012). Late accretion postdating core formation appears to have enriched the silicate Earth in Se as indicated by isotopic data (Varas-Reus et al., 2019). In igneous systems, isotopic fractionations associated with differentiation, sulphide immiscibility or degassing are minimal (<0.5 ‰), such that Se isotopes can act as a fingerprint of the Se source (Kurzawa et al., 2019; Varas-Reus et al., 2019; Yierpan et al., 2019). This approach has revealed isotopic heterogeneities of more than 0.5 ‰ caused by the introduction of sedimentary materials into subduction zones (Kurzawa et al., 2019). Hence Se isotopes in igneous settings are a useful proxy for recycling of volatiles.

3.2 Environmental processes

Most work in Se isotope geochemistry has focused on low-temperature reactions, which can impart large fractionations under certain conditions. Beginning with its liberation from igneous rocks, Se isotopes can be slightly fractionated (ε < 0.5 ‰; Johnson et al., 1999) during oxidative weathering. Much larger isotopic shifts (>10 ‰) have been documented in exceptionally Se-rich soils (up to ~2 wt. % Se; Zhu et al., 2014) and from weathered organic-rich black shales (Clark & Johnson, 2010); however, these conditions are not thought to be representative of the bulk crust, which has much lower Se concentrations (~30 ppb) (Stüeken, 2017) and are therefore unlikely to represent the global Se weathering flux. The settings investigated by Zhu et al. (2014) and Clark & Johnson (2010) are characterized by high concentrations of organic carbon, which may favour re-reduction reactions and isotopic fractionations that are perhaps not typical or organic-lean settings. Importantly, a global mass balance of Se sources and sinks is inconsistent with a significant isotopic fractionation during weathering (Stüeken, 2017), but additional studies are needed to verify this conclusion. In any case, conditions that fractionate Se isotopes during weathering may be locally important, particularly in areas with Se pollution from weathering of exposed Se-rich rocks (see, for example, Stillings & Amacher, 2010).

The Se oxyanions generated during oxidative weathering are transported to the ocean by rivers. The primary fate of these oxyanions on reaching the ocean

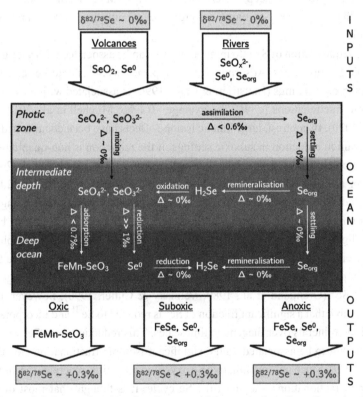

Figure 2 Schematic of major fluxes and isotopic fractionations in the global Se cycle

'Δ' denotes isotopic fractionation associated with a given pathway; '$\delta^{82/78}Se$' denotes isotopic composition of an input or output flux. Darker shading reflects increasing ocean depth.

is uptake into the biomass of phytoplankton (Figure 2). Biological uptake is capable of generating small kinetic isotopic fractionations in algae (<0.6 ‰; Clark & Johnson, 2010), with biomass depleted in heavy isotopes. However, the expressed fractionation may be smaller in the environment – particularly if Se is completely scavenged, as is suggested by its nutrient-type profile in the ocean (Cutter & Cutter, 2001). As biomass sinks from the photic zone to ocean depths, the remineralisation of organic matter releases Se back into seawater in the Se^{-II} oxidation state. This process is not known to fractionate Se isotopes. Throughout most of the modern ocean, liberated Se^{-II} is efficiently oxidized to Se^{IV} (Cutter & Bruland, 1984); anoxic basins are an exception, where Se^{-II} dominates the dissolved Se pool below the chemocline (e.g. Cutter, 1982). While Se^{VI} is the thermodynamically stable form of Se in fully oxic ocean waters, Se^{IV} is kinetically stable, with a residence time in deep waters of $\sim 10^3$ yr (Cutter & Bruland, 1984). Thus, the deep ocean contains a roughly equal mixture of Se^{IV} and Se^{VI} oxyanions.

The sequestration of Se into mineral phases can be associated with a range of isotopic fractionations. Adsorption of Se (typically as Se^{IV}) onto Fe- and Mn-oxides in oxic settings (Balistrieri & Chao, 1990) is associated with only small isotopic fractionations (ε <0.7 ‰, average ~0.1 ‰; Mitchell et al., 2013; Xu et al., 2020). In contrast, large kinetic isotope effects have been documented for Se oxyanion reduction in suboxic settings, if the reduction is non-quantitative. Reduction pathways can be either biotically (ε up to 14 ‰; Herbel et al., 2000; Ellis et al., 2003) or abiotically (ε up to 23 ‰; Johnson & Bullen, 2003) mediated. The magnitude of the fractionation can vary widely, depending on environmental and biological conditions (e.g. Schilling et al., 2020). In both cases, lighter isotopes are preferentially sequestered in the reduced (immobile) phase, either as elemental Se^0 nanoparticles or as Se^{-II} substituted in sulfide mineral phases. In modern reducing sediments, most Se reduction is thought to stop at Se^0 (Oremland et al., 1989; Velinsky & Cutter, 1990); however, it is conceivable that a significant fraction of $Se°$ is reduced to Se^{-II} and incorporated in sulfide minerals over diagenetic timescales. This reduction from Se^0 to Se^{-II} appears not to be associated with a significant isotopic fractionation (Johnson et al., 1999). Since non-quantitative Se oxyanion reduction generates the largest isotopic fractionations known in the Se cycle, it is thought that most of the observed isotopic variability among different depositional environments is related to the variable expression of this fractionation. However, abiotic and biotic reduction pathways can so far not be distinguished with isotopic techniques.

3.3 The Modern Se Cycle

As described above, Se isotope mass balance in the modern ocean is thought to be dominated by the expression of isotope fractionation during oxyanion reduction. Assuming negligible fractionations during weathering and transport, based on global mass-balance calculations (Stüeken, 2017), Se entering the ocean from riverine transport likely has an isotopic composition similar to that of the upper continental crust (~0 ‰; Stüeken, 2017). Input of volatile Se to the ocean via volcanic outgassing is thought to be a minor source (~10%; Stüeken, 2017). Since Se degassing from magmas has negligible isotopic effects (Kurzawa et al., 2019), the isotopic composition of volcanogenic Se probably matches that of the crust (Figure 2).

Non-quantitative Se oxyanion reduction in suboxic bottom waters and sediment pore-waters comprises a sink for light Se isotopes ($\delta^{82/78}$Se < 0). By mass balance, residual dissolved Se (i.e., the marine dissolved Se reservoir) should be isotopically heavier. This is consistent with observations of slightly elevated Se isotope ratios in modern seawater (+0.3 ‰; Chang et al., 2017), FeMn-oxides (+0.3 ‰; Rouxel et al., 2002) and phytoplankton biomass (+0.3 ‰; Mitchell et al., 2012). Oxic sediments are thought to roughly record the isotopic composition of dissolved Se due to removal chiefly via adsorption to FeMn-oxides (which does not substantially fractionate Se isotopes; Figure 2). Anoxic settings are likewise thought to roughly capture the isotopic composition of dissolved Se due to quantitative Se oxyanion reduction (Figure 2). This prediction is analogous to what has been observed for the modern Mo cycle (Kendall et al., 2017).

Removal of Se from the ocean is relatively efficient, yielding a short residence time (~10^3–10^4 yr) for dissolved Se in the deep ocean (Cutter & Bruland, 1984; Stüeken, 2017). As this residence time is of similar magnitude to the ocean mixing time, Se is only moderately well mixed throughout the global ocean. As such, paleo-redox reconstructions utilizing Se abundances or isotopic ratios are sensitive to only local, perhaps basin-scale redox processes, which may mask global trends. Sediments that are depleted in heavy Se isotopes relative to seawater ($\delta^{82/78}$Se < +0.3) likely reflect deposition under a sizable Se oxyanion reservoir, such that reductive immobilization in bottom waters or sediment porewaters was non-quantitative. Such non-quantitative reduction is observed in modern marine sediments underlying oxic-to-suboxic waters (Johnson, 2004). In contrast, sediments that record seawater Se isotope ratios ($\delta^{82/78}$Se ~ +0.3) have two possible explanations: if organic poor, then Se may be present predominantly as adsorbed to FeMn-oxides under fully oxic conditions; if organic rich, the unfractionated Se isotopes are more likely indicative of quantitative Se reduction under anoxic diagenetic conditions. Quantification of the Se held in different host phases can distinguish these two scenarios.

In order for sediments to become enriched in heavy Se isotopes ($\delta^{82/78}$Se > +0.3), non-quantitative reduction must be operating prior to influx of Se into the basin, such that lighter Se isotopes are sequestered elsewhere and the residual pool of Se oxyanions becomes isotopically enriched. Such scenarios can be encountered in redox-stratified water masses (where Se can be sequestered in shallower sediments; Stüeken et al., 2015a; Kipp et al., 2017) or potentially in restricted basins that receive Se from upwelling (where Se is sequestered nearer to the upwelling source; Kipp et al., 2020).

It is important to note that while the direction of these fractionations is likely applicable to deep-time studies, the inflection point (i.e., seawater Se isotope ratio) is likely to have changed as a function of the relative magnitude of oxic/suboxic/anoxic sinks over geologic time. Furthermore, the isotopic composition of seawater may be spatially variable, depending on biological productivity and Se supply from terrestrial sources. These uncertainties affect the types of research question that can be addressed with this proxy. Se isotopes are not as powerful as Mo isotopes for reconstructing the long-term evolution of the global ocean from a few selected study sites (Kendall et al., 2017); however, data from individual sites can reveal whether or not surface environments in the past were capable of generating and sustaining a significant a Se oxyanion reservoir, which makes Se isotopes useful for tracking the early redox history of our planet (see Section 4). Furthermore, the fact that Se isotopes are sensitive to regional environmental factors makes them a powerful tool for mapping out redox heterogeneities in deep time.

4 Case Study

To illustrate the utility of Se as a paleo-redox proxy, we consider two well-studied packages of black shales that were deposited in the late Archean, just prior to the first permanent rise of atmospheric oxygen during the Great Oxidation Event. Several studies over the past two decades have implicated transient, spatially limited oxygenation of surface environments during the late Archean. Among the best known of these events is the 'whiff' of oxygen recorded in the 2.5 Ga Mt. McRae shale (Anbar et al., 2007), which was initially recognized on the basis of Mo and Re enrichments that are indicative of oxidative weathering. Accompanying the Mo enrichment is a positive excursion in nitrogen isotope ratios (δ^{15}N) suggestive of local aerobic nitrogen cycling (Garvin et al., 2009), which implicates an expansion of oxygenated surface waters following the pulse of oxidative weathering.

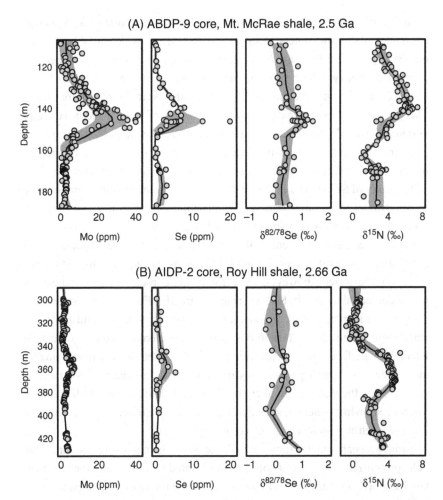

Figure 3 Chemostratigraphy of the (A) 2.5 Ga Mt. McRae shale and (B) 2.66 Ga Roy Hill shale. Both of these units record geochemical evidence for transient marine oxygenation. The analysis of Se contents and isotopic composition allows a more precise comparison to be drawn between these events, revealing that the Mt. McRae 'whiff' of oxygen evidently involved a larger magnitude of oxidative weathering and ocean oxygenation than the Roy Hill event.

In order to gain more detailed insight into the processes occurring during this 'whiff' of atmospheric oxygen, Stüeken et al. (2015a) measured Se concentrations and isotope ratios along the same drill core. The data revealed a pulse of Se enrichment in tandem with the Mo spike (Figure 3A). Furthermore, Se isotopes showed a positive excursion at the peak of the 'whiff' interval (Figure 3A). This isotopic enrichment was interpreted as a signal of non-quantitative Se reduction occurring in shallow waters along the basin margin, which may have been mildly oxygenated. By mass balance, the residual Se oxyanion pool became isotopically heavy, which was recorded via quantitative Se reduction in the anoxic deeper facies, including the black shales that were sampled in that study. This Se isotope trend reveals an important aspect of basinal redox structure during the 'whiff' interval: deep waters did not become oxygenated, in spite of concurrent evidence for atmospheric and surface water oxygenation, but marine margins were sufficiently oxic to maintain a small Se oxyanion reservoir.

The study of Se isotopes in the Mt. McRae shale gains additional importance by comparison with other transient oxygenation events in the Late Archean ocean. Recent work by Koehler et al. (2018) and Olson et al. (2019) revealed a transient oxygenation event in the 2.66 Ga Roy Hill shale of the Jeerinah Formation. This work revealed a similar δ^{15} N excursion, indicative of oxygenated surface waters; however, sedimentary enrichments of Mo and Se are muted compared to the Mt. McRae shale (Figure 3B). Furthermore, Se isotopes show no clear trend across the interval of the δ^{15} N excursion and muted Se enrichment (Figure 3B). Thus, in this case, it can be concluded that a transient interval of surface water oxygenation was associated with a much smaller magnitude of atmospheric oxygenation and oxidative weathering. Additionally, the chemocline likely persisted at a shallower depth than during Mt. McRae 'whiff' interval, such that Mo and Se were efficiently scavenged in shallower settings on reaching the ocean.

Surface ocean oxygenation on a large scale is not expressed until the Paleoproterozoic where Se isotopes show elevated values in multiple sedimentary basins (Kipp et al., 2017) (Figure 4). Evidence for Se oxyanions penetrating into the deep ocean becomes more widespread in the late Neoproterozoic and Phanerozoic (Pogge von Strandmann et al., 2015; Stüeken et al., 2015b; Mitchell et al., 2016), but more detailed analyses of multiple basins at higher temporal and spatial resolution are required to assess how widely these signatures are expressed. Using these two case studies from the Neoarchean as a guide, future paired studies of Se with proxies for local surface water oxygenation (e.g., N isotopes) and global ocean redox (e.g., Mo and U isotopes) have the potential to reveal nuanced information about the temporal and spatial extent of redox fluctuations in ancient environments.

5 Future Prospects

Se isotopes have great potential to provide additional insights into redox perturbations over Earth's history, in particular at the upper end of the Eh

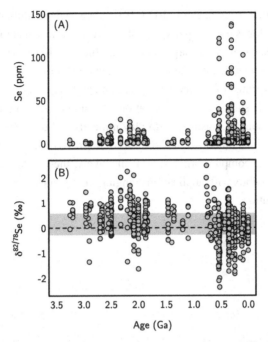

Figure 4 Secular evolution of (A) Se concentration and (B) Se isotopic composition in siliciclastic marine sedimentary rocks. Increases in Se enrichment at the beginning and end of the Proterozoic Eon correspond to Earth's major oxygenation events. In tandem, Se isotopes experience large fluctuations at these times, reflecting a shift in the prevalence of Se oxyanion reduction. Data were compiled in Stüeken et al. (2015b), Mitchell et al. (2016) and Kipp et al. (2017, 2020).

scale where other proxies such as S or Mo isotopes are less sensitive. However, the development of the proxy is still in its early stages, and additional work is needed to extract more quantitative information from the rock record. First, as noted earlier, the future of Se isotopes as a biogeochemical proxy may lie in phase-specific analyses such that isotopic signals archived in distinct mineral phases and organic matter can be disentangled. Second, the large number of stable isotopes in the Se system invites closer inspection of mass-independent fractionation (MIF) pathways. Existing data do not show any evidence for MIF; however, most studies have focused on masses 76, 78 and 82. It is conceivable that abundance-dependent fractionation exists for masses 74 and 80. Alternatively, mass 77 may show a distinct behaviour as it is the only odd-numbered isotope. It is also possible that the proportionality between the difference masses changes as a function of the reaction pathway as documented for sulphur isotopes (Johnston, 2011). If so, then high-resolution analyses of multiple Se isotopes may help distinguish between biotic and abiotic reactions, for example. Such analyses may become possible with future developments in collision cell technologies. Lastly, more experimental work is needed to develop fingerprints for numerous environmental reactions that are so far uncharacterized, such as photo-oxidation, oxyanion incorporation into sulphates and carbonates, adsorption to clay minerals and organics or biological fractionations at nanomolar concentrations of dissolved Se, including Se metabolisms in more complex organisms. Collectively, these data will enrich our geochemical toolbox with applications ranging from planetary science to environmental pollution.

References

Anbar A, Duan Y, Lyons TW, Arnold GL, Kendall B, Creaser RA, Kaufman AJ, Gordon GW, Scott CT, Garvin J and Buick R (2007) A whiff of oxygen before the Great Oxidation Event? *Science*, **317**, 1903–6.

Balistrieri LS and Chao TT (1990) Adsorption of selenium by amorphous iron oxyhydroxide and manganese dioxide. *Geochimica et Cosmochimica Acta*, **54**, 739–51.

Banning H, Stelling M, König S, Schoenberg R and Neumann T (2018) Preparation and purification of organic samples for selenium isotope studies. *PloS ONE*, **13**, doi:10.1371/journal.pone.0193826.

Chang Y, Zhang J, Qu JQ and Xue Y (2017) Precise selenium isotope measurement in seawater by carbon-containing hydride generation-desolvation-MC-ICP-MS after thiol resin preconcentration. *Chemical Geology*, **471**, 65–73.

Clark SK and Johnson TM (2010) Selenium stable isotope investigation into selenium biogeochemical cycling in a lacustrine environment: Sweitzer Lake, Colorado. *Journal of Environmental Quality*, **39**, 2200–10.

Cutter GA (1982) Selenium in reducing waters. *Science*, **217**, 829–31.

Cutter GA and Bruland KW (1984) The marine biogeochemistry of selenium: a re-evaluation. *Limnology and Oceanography*, **29**, 1179–92.

Cutter GA and Cutter LS (2001) Sources and cycling of selenium in the western and equatorial Atlantic Ocean. *Deep-Sea Research II*, **48**, 2917–31.

Ellis AS, Johnson TM, Herbel MJ and Bullen T (2003) Stable isotope fractionation of selenium by natural microbial consortia. *Chemical Geology*, **195**, 119–29.

Fan H, Wen H, Hu R and Zhao H (2011) Selenium speciation in Lower Cambrian Se-enriched strata in South China and its geological implications. *Geochimica et Cosmochimica Acta*, **75**, 7725–40.

Garvin J, Buick R, Anbar AD, Arnold GL and Kaufman AJ (2009) Isotopic evidence for an aerobic nitrogen cycle in the latest Archean. *Science*, **323**, 1045–8.

Gladyshev VN (2012) Selenoproteins and selenoproteomes. In: *Selenium: its molecular biology and role in human health* (eds. Hatfield DL, Berry MJ and Gladyshev VN). Springer Science + Business Media LLC, pp. 109–23.

Goldschmidt VM and Hefter O (1933) Zur Geochemie des Selens. *Nachrichten von der Gesellschaft der Wissenschaften zu Goettingen. Fachgruppe III*, **35–6**, 245–51.

Herbel MJ, Johnson TM, Oremland RS and Bullen T (2000) Fractionation of selenium isotopes during bacterial respiratory reduction of selenium oxyanions. *Geochimica et Cosmochimica Acta*, **64**, 3701–9.

Johnson TM (2004) A review of mass-dependent fractionation of selenium isotopes and implications for other heavy stable isotopes. *Chemical Geology*, **204**, 201–14.

Johnson TM and Bullen T (2003) Selenium isotope fractionation during reduction by Fe(II)-Fe(III) hydroxide-sulfate (green rust). *Geochimica et Cosmochimica Acta*, **67**, 413–9.

Johnson TM, Herbel MJ, Bullen TD and Zawislanski PT (1999) Selenium isotope ratios as indicators of selenium sources and oxyanion reduction. *Geochimica et Cosmochimica Acta*, **63**, 2775–83.

Johnston DT (2011) Multiple sulfur isotopes and the evolution of Earth's surface sulfur cycle. *Earth-Science Reviews*, **106**.

Kendall B, Dahl TW and Anbar AD (2017) The stable isotope geochemistry of molybdenum. *Reviews in Mineralogy and Geochemistry*, **82**, 683–732.

Kipp MA, Algeo TJ, Stüeken EE and Buick R (2020) Basinal hydrographic and redox controls on selenium enrichment and isotopic composition in Paleozoic black shales. *Geochimica et Cosmochimica Acta*, doi:10.1016/j.gca.2019.1012.1016.

Kipp MA, Stüeken EE, Bekker A and Buick R (2017) Selenium isotopes record extensive marine suboxia during the Great Oxidation Event. *Proceedings of the National Academy of Sciences*, **114**, 875–80.

Koehler MC, Buick R, Kipp MA, Stüeken EE and Zaloumis J (2018) Transient surface oxygenation recorded in the ~2.66 Ga Jeerinah Formation, Australia. *Proceedings of the National Academy of Sciences*, **115**, 7711–16.

König S, Luguet A, Lorand JP, Wombacher F and Lissner M (2012) Selenium and tellurium systematics of the Earth's mantle from high precision analyses of ultra-depleted orogenic peridotites. *Geochimica et Cosmochimica Acta*, **86**, 354–66.

Krouse HR and Thode HG (1962) Thermodynamic properties and geochemistry of isotopic compounds of selenium. *Canadian Journal of Chemistry*, **40**, 367–75.

Kurzawa T, König S, Alt JC, Yierpan A and Schoenberg R (2019) The role of subduction recycling on the selenium isotope signature of the mantle: constraints from Mariana arc lavas. *Chemical Geology*, **513**, 239–49.

Kurzawa T, König S, Labidi J, Yierpan A and Schoenberg R (2017) A method for Se isotope analysis of low ng-level geological samples via double spike and hydride generation MC-ICP-MS. *Chemical Geology*, **466**.

Labidi J, König S, Kurzawa T, Yierpan A and Schoenberg R (2018) The selenium isotopic variations in chondrites are mass-dependent: implications for sulfide formation in the early solar system. *Earth and Planetary Science Letters*, **481**, 212–22.

Large RR, Halpin JA, Danyushevsky LV, Maslennikov VV, Bull SW, Long JA, Gregory DD, Lounejeva E, Lyons TW, Sack PJ, McGoldrick JJ and Calver CR (2014) Trace element content of sedimentary pyrite as a new proxy for deep-time ocean-atmosphere evolution. *Earth and Planetary Science Letters*, **389**, 209–20.

Mitchell CE, Couture R-M, Johnson TM, Mason PRD and Van Cappellen P (2013) Selenium sorption and isotope fractionation: iron(III) oxides versus iron(II) sulfides. *Chemical Geology*, **342**, 21–8.

Mitchell K, Mansoor SZ, Mason PR, Johnson TM and Van Cappellen P (2016) Geological evolution of the marine selenium cycle: Insights from the bulk shale $\delta^{82/76}$Se record and isotope mass balance modeling. *Earth and Planetary Science Letters*, **441**, 178–87.

Mitchell K, Mason PRD, Van Cappellen P, Johnson TM, Gill BC, Owens JD, Ingall ED, Reichart G-J and Lyons TW (2012) Selenium as paleo-oceanographic proxy: a first assessment. *Geochimica et Cosmochimica Acta*, **89**, 302–17.

Olson SL, Ostrander CM, Gregory DD, Roy M, Anbar AD and Lyons TW (2019) Volcanically modulated pyrite burial and ocean–atmosphere oxidation. *Earth and Planetary Science Letters*, **506**, 417–27.

Oremland RS (1990) Measurement of in situ rates of selenate removal by dissimilatory bacterial reduction in sediments. *Environmental Science and Technology*, **24**, 1157–64.

Oremland RS, Hollibaugh JT, Maest AS, Presser TS, Miller LG and Culbertson CW (1989) Selenate reduction to elemental selenium by anaerobic bacteria in sediments and culture: biogeochemical significance of a novel, sulfate-independent respiration. *Applied and Environmental Microbiology*, **55**, 2333–43.

Pogge von Strandmann PA, Coath CD, Catling DC, Poulton SW and Elliott T (2014) Analysis of mass dependent and mass independent selenium isotope variability in black shales. *Journal of Analytical Atomic Spectrometry*, **29**, 1648–59.

Pogge von Strandmann PA, Stüeken EE, Elliott T, Poulton SW, Dehler CM, Canfield DE and Catling DC (2015) Selenium isotope evidence for progressive oxidation of the Neoproterozoic biosphere. *Nature Communications*, **6**, doi:10.1038/ncomms10157.

Pons ML, Millet MA, Nowell GN, Misra S and Williams HM (2020) Precise measurement of selenium isotopes by HG-MC-ICPMS using a 76–78 double-spike. *Journal of Analytical Atomic Spectrometry*, **35**.

Rouxel O, Ludden J, Carignan J, Marin L and Fouquet Y (2002) Natural variations of Se isotopic composition determined by hydride generation multiple collector inductively coupled plasma mass spectrometry. *Geochimica et Cosmochimica Acta*, **66**, 3191–9.

Rue EL, Smith GJ, Cutter GA and Bruland KW (1997) The response of trace element redox couples to suboxic conditions in the water column. *Deep Sea Research Part I: Oceanographic Research Papers*, **44**, 113–34.

Schilling K, Basu A, Wanner C, Sanford RA, Pallud C, Johnson TM and Mason PR (2020) Mass-dependent selenium isotopic fractionation during microbial reduction of seleno-oxyanions by phylogenetically diverse bacteria. *Geochimica et Cosmochimica Acta*, doi:10.1016/j.gca.2020.1002.1036.

Schilling K, Johnson TM, Dhillon KS and Mason PR (2015) Fate of selenium in soils at a seleniferous site recorded by high precision Se isotope measurements. *Environmental Science & Technology*, **49**, 9690–8.

Schilling K, Johnson TM and Mason PR (2014) A sequential extraction technique for mass-balanced stable selenium isotope analysis of soil samples. *Chemical Geology*, **381**, 125–30.

Stillings LL and Amacher MC (2010) Kinetics of selenium release in mine waste from the Meade Peak phosphatic shale, Phosphoria formation, Wooley Valley, Idaho, USA. *Chemical Geology*, **269**, 113–23.

Stolz JF, Basu P, Santini JM and Oremland RS (2006) Arsenic and selenium in microbial metabolism. *Annual Review in Microbiology*, **60**, 107–30.

Stüeken EE (2017) Selenium isotopes as a biogeochemical proxy in deep time. *Reviews in Mineralogy and Geochemistry*, **82**, 657–82.

Stüeken EE, Buick R and Anbar AD (2015a) Selenium isotopes support free O_2 in the latest Archean. *Geology*, **43**, 259–62.

Stüeken EE, Buick R, Bekker A, Catling D, Foriel J, Guy BM, Kah LC, Machel HG, Montañez IP and Poulton SW (2015b) The evolution of the global selenium cycle: secular trends in Se isotopes and abundances. *Geochimica et Cosmochimica Acta*, **162**, 109–25.

Stüeken EE, Foriel J, Buick R and Schoepfer SD (2015c) Selenium isotope ratios, redox changes and biological productivity across the end-Permian mass extinction. *Chemical Geology*, **410**, 28–39.

Stüeken EE, Foriel J, Nelson BK, Buick R and Catling DC (2013) Selenium isotope analysis of organic-rich shales: advances in sample preparation and isobaric interference correction. *Journal of Analytical Atomic Spectrometry*, **28**, 1734–49.

Varas-Reus MI, König S, Yierpan A, Lorand JP and Schoenberg R (2019) Selenium isotopes as tracers of a late volatile contribution to Earth from the outer Solar System. *Nature Geoscience*, **12**, 779–82.

Velinsky DJ and Cutter GA (1990) Determination of elemental selenium and pyrite-selenium in sediments. *Analytica Chimica Acta*, **235**, 419–25.

Wachsmann M and Heumann KG (1992) Negative thermal ionization mass spectrometry of main group elements Part 2. 6th group: sulfur, selenium and tellurium. *International Journal of Mass Spectrometry and Ion Processes*, **114**, 209–20.

Wen H and Carignan J (2007) Reviews on atmospheric selenium: emissions, speciation and fate. *Atmospheric Environment*, **41**, 7151–65.

Xu W, Zhu JM, Johnson TM, Wang X, Lin ZQ, Tan D and Qin H (2020) Selenium isotope fractionation during adsorption by Fe, Mn and Al oxides. *Geochimica et Cosmochimica Acta*, **272**, 121–36.

Yierpan A, König S, Labidi J and Schoenberg R (2019) Selenium isotope and S-Se-Te elemental systematics along the Pacific-Antarctic ridge: role of mantle processes. *Geochimica et Cosmochimica Acta*, **249**, 199–224.

Zhu JM, Johnson TM, Clark SK, Zhu XK and Wang XL (2014) Selenium redox cycling during weathering of Se-rich shales: A selenium isotope study. *Geochimica et Cosmochimica Acta*, **126**, 228–49.

Cambridge Elements ≡

Elements in Geochemical Tracers in Earth System Science

Timothy Lyons

University of California

Timothy Lyons is a Distinguished Professor of Biogeochemistry in the Department of Earth Sciences at the University of California, Riverside. He is an expert in the use of geochemical tracers for applications in astrobiology, geobiology and Earth history. Professor Lyons leads the 'Alternative Earths' team of the NASA Astrobiology Institute and the Alternative Earths Astrobiology Center at UC Riverside.

Alexandra Turchyn

University of Cambridge

Alexandra Turchyn is a University Reader in Biogeochemistry in the Department of Earth Sciences at the University of Cambridge. Her primary research interests are in isotope geochemistry and the application of geochemistry to interrogate modern and past environments.

Chris Reinhard

Georgia Institute of Technology

Chris Reinhard is an Assistant Professor in the Department of Earth and Atmospheric Sciences at the Georgia Institute of Technology. His research focuses on biogeochemistry and paleoclimatology, and he is an Institutional PI on the 'Alternative Earths' team of the NASA Astrobiology Institute.

About the series

This innovative series provides authoritative, concise overviews of the many novel isotope and elemental systems that can be used as 'proxies' or 'geochemical tracers' to reconstruct past environments over thousands to millions to billions of years – from the evolving chemistry of the atmosphere and oceans to their cause-and-effect relationships with life.

Covering a wide variety of geochemical tracers, the series reviews each method in terms of the geochemical underpinnings, the promises and pitfalls, and the 'state-of-the-art' and future prospects, providing a dynamic reference resource for graduate students, researchers and scientists in geochemistry, astrobiology, paleontology, paleoceanography and paleoclimatology.

The short, timely, broadly accessible papers provide much-needed primers for a wide audience – highlighting the cutting edge of both new and established proxies as applied to diverse questions about Earth system evolution over wide-ranging time scales.

Elements in Geochemical Tracers in Earth System Science

Elements in the series

Triple Oxygen Isotopes
Huiming Bao

The Uranium Isotope Paleoredox Proxy
Kimberly V. Lau et al.

Application of Thallium Isotopes
Jeremy D. Owens

Selenium Isotope Paleobiogeochemistry
Eva E. Stüeken and Michael A. Kipp

The Pyrite Trace Element Paleo-Ocean Chemistry Proxy
Daniel D. Gregory

Earth History of Oxygen and the iprOxy
Zunli Lu, Wanyi Lu, Rosalind E. M. Rickaby and Ellen Thomas

A full series listing is available at: www.cambridge.org/EESS